32d Congress, 1st Session. [SENATE.] Ex. Doc. No. 81.

REPORT

OF

THE SECRETARY OF WAR,

COMMUNICATING,

In compliance with a resolution of the Senate, a reconnoissance of the Gulf of California and the Colorado river by Lieutenant Derby.

June 19, 1852.

Ordered to lie on the table and be printed.

War Department,
Washington, June 15, 1852.

Sir: In compliance with the resolution of the Senate of the 10th ultimo, I have the honor to transmit herewith a report of the colonel of topographical engineers, with a copy of the reconnoissance of the Gulf of California and the Colorado river, made in 1850 and 1851 by Lieutenant Derby, of the topographical engineers.

The report of the reconnoissance of Tulare lake, also called for by the resolution, has not yet been received at this department.

Very respectfully, your obedient servant,

C. M. CONRAD,
Secretary of War.

Hon. Wm. R. King,
President of the Senate.

Bureau of Topographical Engineers,
Washington, June 14, 1852.

Sir: A resolution of the Senate of the 10th of May, calls for a copy "of the reconnoissance of the Gulf of California and the Colorado river made in 1850-'51, by Lieutenant Derby, of the corps of topographical engineers, together with a copy of the report made by said officer to the department in relation to the Tulare lake and its vicinity."

In obedience I have the honor to submit a copy of the reconnoissance of the Gulf of California and of the Colorado river; but it is not in my power to send a copy of the report of the reconnoissance of Tulare lake, as it has not been received at this, or by my inquiries at any other office in this city.

Respectfully, sir, your obedient servant,

J. J. ABERT,
Colonel corps Topographical Engineers.

Hon. C. M. Conrad,
Secretary of War.

Report of the Expedition of the United States Transport "Invincible," (Capt. A. H. Wilcox,) made by order of Major General P. F. Smith, commanding Pacific division, to the Gulf of California and river Colorado, during the months of November, December, January, February and March, 1850 and 1851. By Geo. H. Derby, Brevet 1st Lieut., Topographical Engineers.

HEADQUARTERS, THIRD DIVISION,
SONOMA, *October* 11, 1850.

SIR: The schooner Invincible is to be despatched by the chief assistant quartermaster of the division to attempt the entrance of the river Colorado from the Gulf of California, and the ascent of the river as far as the post to be established near the junction of the Gila, with a view to establish that as a route of supply.

As this will afford a good opportunity of determining with accuracy many points in the geography of that part of the country of great future consequence, you will proceed to San Diego to embark in the vessel that may be sent on that service.

You will fix by observations and deductions as far as possible all of the points about the head of the Gulf of California important to the navigation of the Colorado. You will record the soundings when they may have been procured, especially at the entrance of the river and in its course upward. You will also ascertain as far as practicable the rise and fall of the tide, and the time of high water at the full and change of the moon at the mouth of the river. Likewise record the character and topography of the country at the head of the gulf and the river, as far as falls within your observation, and make a sketch of all the coast and shores in that neighborhood; and procure all such other information as may be of use to the service and the public generally.

The principal object of the expedition is to open a route of transportation by water to the mouth of the Gila, for the supply of the post to be established there. There are no means for an accurate or extensive survey, but the opportunity of gathering much useful information must not be lost, and a skilful and industrious officer is needed for that purpose. But your movements must be subordinate to those of the vessel, and her principal object must never be abandoned for the sake of even the most interesting results not connected with it.

Instructions will be given, however, to the quartermaster's department, that will insure every assistance and coöperation that can be furnished without detriment to the main end in view.

As no military escort can be furnished, no risk will be run in those sections infested by hostile Indians.

Very respectfully, your obedient servant,

J. HOOKER,
Assistant Adjutant General.

Lieut. G. H. DERBY,
Topographical Engineers, Sonoma.

SONOMA, CALIFORNIA,
March 31, 1851.

COLONEL: In compliance with your instructions I left Sonoma, arriving at Benicia on the 12th of October ultimo, where I had an interview with

Major Robert Allen, the chief quartermaster of the division, who supplied me with funds for any necessary expenditure that might occur during the expedition.

I proceeded to San Francisco on the 13th, intending to take passage immediately for San Diego, but finding that there was no steamer to sail upon the 15th, and there being no other method of transportation available, I was compelled to remain until the arrival of the schooner Invincible, which took place on the 31st.

I had taken from the topographical office at division headquarters, one chronometer, (Bliss & Creighton, No. 1468,) two sextants, and two Schmalcalder's compasses, and was accompanied by Mr. Walter H. Ferrier, a young gentleman who had been employed to assist me in making astronomical observations, and in such surveys as might be practicable. I embraced the opportunity of the unavoidable delay at San Francisco to have my sextants thoroughly adjusted, and placed my chronometer with Captain Alexander Frazer, of the revenue service, who had established an observatory at Rincon point for rating those instruments. He discovered in a few days that it was working very irregularly, owing to some internal derangement, and I was obliged to have it removed and repaired. As I had no time to have it satisfactorily rated after its repair, and had an opportunity to procure an excellent instrument at a reasonable price, I purchased chronometer No. 1430, Parkinson and Frodsham, which has performed with the most satisfactory results during the expedition.

The transport Invincible, Captain A. H. Wilcox, which had been designated for the service, is a top-sail schooner of 120 tons burden, and draws when light between eight and nine feet water. She had a crew of nine seamen, cook, steward and two mates, and was supplied with provisions for a four months' cruise. An old iron twelve-pounder carronade (captured from the Mexicans at Tabasco) and six muskets with flint locks comprised the extent of her armament.

Everything being ready for our departure, we sailed from San Francisco on the 1st November with a fine breeze from the northwest, before which we rapidly ran down the coast, and passing the island of Santa Barbara upon the 2d, and St. Clements upon the 3d, we came to anchor upon the 4th at 11 a. m., off the beach or "Plaza" of San Diego. Here the vessel was detained until the 15th discharging cargo, taking in ballast and repairing her copper, during which interval I obtained a copy of Imray's chart of the Gulf of California and Lieutenant Hardy's "Travels in Mexico," which last, though very inaccurate in many important points, was nevertheless of much assistance to us in the navigation of the river. I had a conversation also with Colonel Williams, an old and well known resident of Los Angelos, who stated that he had explored the banks of the Colorado nearly to its entrance into the gulf, and gave me such fearful accounts of the ferocity of the Indians inhabiting the country in the vicinity of the river, that I applied to Lieutenant Colonel Magruder, commanding the light battery stationed at San Diego, for some addition to our armament, which, however, he was unable to supply. Having received on board ten thousand rations for the command stationed at the junction of the Gila and Colorado, we sailed from San Diego on the afternoon of the 15th, and on the 17th passed the island of Guadaloupe, situated in latitude 29° 20′.

This island is about fifteen miles in length and five in width. It is rocky and mountainous, but covered with vegetation, and is represented to be thickly inhabited by wild goats of unusual size. Water is found upon the eastern shore, and the island is frequently visited by small vessels engaged in the capture of the sea-elephant, numbers of which animals are found upon its coast. As we passed it at a distance of thirty miles, I was unable to make a sketch of its outline. On the 19th we passed the Alijos rocks, bearing S.S.W. and distant about ten miles. This remarkable reef, first discovered in 1710, is about four miles in length, the highest rock being about 110 feet above the surface. Its general direction is N.E. and S.W. The entire reef is so thickly covered with guano that we found it impossible to form any conjecture as to its geological character. The surf dashing over it in a high sea is a magnificent spectacle when beheld at a safe distance, but the vigias surrounding would render a close view dangerous. I believe no one has ever attempted a landing. The wind continuing favorable, we made the southern extremity of the peninsula of California upon the 22d at noon, Cape Falso bearing N. by W. and Cape St. Lucas N.E. by E. at about fifteen miles distance. The land is high, bold, and in the vicinity of the capes, mountainous. The white chalky cliffs which may be seen at a great distance render the coast easy of recognition.

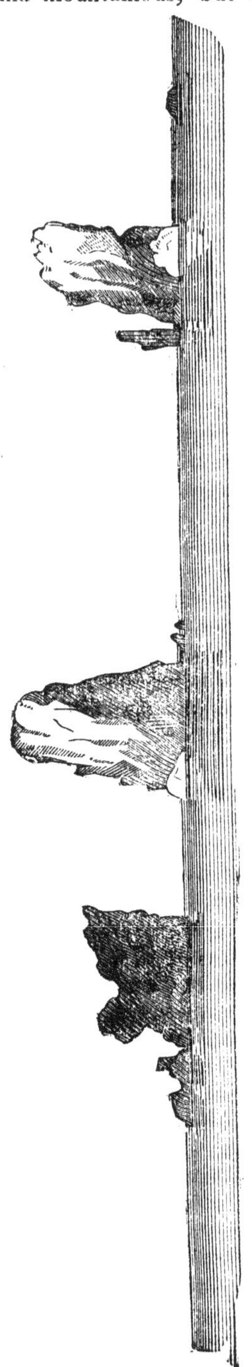

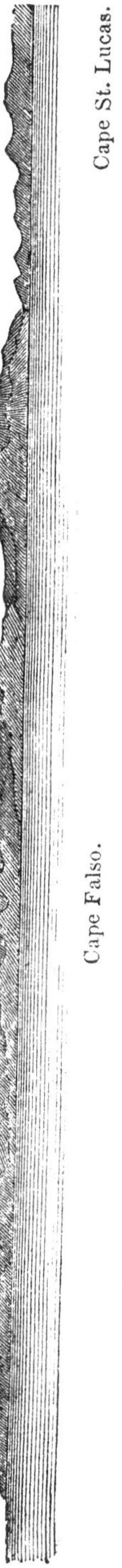

Cape St. Lucas.

Cape Falso.

We ran into the little harbor of Cape St. Lucas on the 23d, anchoring in sixteen fathoms water, within a cable's length of the shore. The harbor is but two miles in width, protected from the southeast by a high rocky bluff, but perfectly exposed to the north and west. It is an extremely dangerous anchorage, as a heavy surf is continually rolling in, and no bottom is found at a hundred fathoms, until close into the beach.

The wreck of the brig Friendship was lying upon the shore, which vessel was beached here three or four months before our arrival. We landed through the surf and walked up to the mission, situated in a little valley a mile from the beach, and consisting of some eight or ten adobe houses thatched with straw. The valley appeared to have been under cultivation, but the inhabitants had nearly all left for the interior to avoid the cholera then raging upon the coast. I observed here a fine herd of cattle, and many pigs, goats and fowls. The number of inhabitants I was informed was about fifty. With the exception of the little valley under cultivation, which is watered by a small stream running from the hills and emptying into the sea at this point, the country is mountainous and sterile, and covered with various species of cacti, which grow to an immense size. Returning to the vessel I noticed upon the beach a large *copper* anchor weighing probably eighteen hundred weight; it was evidently a vestage of antiquity, being singularly shaped, with a remarkable long shank and narrow flukes. I was informed by an American resident of the place (Mr. Wilks) that it was fished up in the harbor three or four years ago. It was probably lost here by some of the early Spanish explorers of California. Leaving Cape St. Lucas with the evening land breeze, we passed Cape Palmo on the 24th, bearing west about thirty miles distant, and on the 25th at noon passed within ten miles of the small island of San Ignacio in latitude 23° 39′ 30″. This island, situated on the eastern side of the gulf, is a barron rock, and entirely uninhabited; it is about five miles long and two or three in width.

I. San Ignacio, bearing N. 30° E. 5 miles.

We experienced a heavy gale from the S. W. commencing on the 26th, which gradually wore round to the N. W. and brought us off the entrance to the harbor of Guyamas on the 28th. We succeeded in entering the harbor on the 29th, where we anchored in four and a half fathoms water, with the intention of remaining a few days to obtain the rates of the chronometers, and take in a supply of water and fresh provisions. I hoped also to obtain some information regarding the navigation of the gulf, from the residents of Guyamas engaged in the pearl fishery, and other persons likely to be acquainted with it.

The harbor of Guyamas is one of the best upon the coast; it is perfectly land locked and protected by its numerous islands from every quarter. It has been so often and so well surveyed, that it is unnecessary for me to enter into a minute description of its merits. It is to be regretted, however, that it has not more water, from four to five fathoms being the average depth, which is insufficient for a ship of the line, or even a first class frigate. We found Guyamas a dirty place, with a dirty population of about 1,500 or 2,000. The houses being built of adobe with the roofs sloping towards the interior, have a very unfinished appearance, and from the harbor the town presents the appearance of having been abandoned when half built. There are two small piers in a ruined condition; and near the landing a large pile of earth, surmounted by two or three crumbling walls, over which floats from a lofty staff the flag of Mexico, marks the site of the adobe fort knocked down by the guns of the Dale during the late war. There are several wealthy individuals in Guyamas, who monopolize the whole of the business with the interior of Sonora, but the mass of the population are in a state of wretched poverty. One or two small vessels from San Francisco were lying in the harbor, the proprietors of which had purchased everything in the shape of fresh provisions to be obtained. Sheep in large numbers they had purchased from the interior of Sonora with the intention of landing them at Molexe, on the California coast, thence driving them overland to San Francisco. Fowls, turkeys, ducks, goats, everything that could be purchased for one real and sold for twenty had disappeared, causing us no little difficulty in obtaining fresh provisions, even at comparatively high prices. The water at Guyamas is obtained from wells, and is slightly brackish. Excellent oysters are brought from the river Yaqui, which empties into the bay about twenty miles south of the town, and sold to the shipping at a dollar a bushel; the Mexicans, however, make no use of them. We found the cholera raging at Guyamas on our arrival, many deaths occurring daily. It appeared to be confined to the native population entirely, which, considering their dirty habits and mode of living, was by no means to be wondered at. Dr. W. L. White, an excellent physician from San Francisco, had recently arrived, and was engaged in a very successful practice. We obtained from him such medicines as were deemed necessary under the circumstances, but fortunately had no occasion to use them. I commenced a series of daily observations upon one of the small islands in the harbor, for the rates of the chronometers, which we succeeded in ascertaining with accuracy.

Immediately upon our arrival I called upon the American consul, Mr. John Robinson, who treated us with civility, and M. Cubillas, the consul of France, who received us with great attention and hospitality, and took much interest in the expedition. I also visited Mr. Thomas Spence, an old English resident, who was the agent of Lieutenant Hardy for the purchase of the "Bruja" for his expedition in 1829, and who had frequently made trips towards the head of the gulf himself, though never going far north of the island of Tiburon. He very kindly gave me all the information in his power in regard to the navigation of the gulf and the small harbors upon the coast north of Guyamas. All of these gentlemen united in saying that the country near the head of the gulf and bordering the river Colorado was inhabited by hostile Indians, from whom we might expect much trouble; and on becoming acquainted with the weakness of our armament, strenuously advised an increase of that and in the number of our crew. I therefore purchased from the captain of a Spanish schooner lying in the harbor two small swivels and eight carbines, with the necessary ammunition, and the captain engaged three additional seamen, increasing the number of our crew to twelve. This force I deemed sufficient for any emergency that might arise.

The time of high water at "full and change" of the moon at Guyamas is not very accurately determined; it is, however, between 8 and 9 a. m. The ordinary rise of spring tides is six feet, neap tides four feet. The phenomena of four tides in twenty-four hours has repeatedly occurred here, as I am credibly informed. The prevailing winds in May, June and July are from the southeast and southwest. The thermometer during the summer months ranges from 92° to 98° Fahrenheit, the maximum 110°; during winter from 56° to 60°, minimum 45°. Guyamas is in latitude 27° 54′ N., longitude 110° 49′ 10″ W.

I. Tortuga, S. 6° W. 10 miles.

We sailed from Guyamas on Friday, the 6th December, and for three days experienced a very heavy gale from the northeast, against which the vessel struggled under double-reefed sails without making much progress to the northward. We passed the small islands of Rolasco, situated about twenty miles west of Guyamas, on the 6th, and on the 7th were within ten miles of the island of Tortuga, bearing from us S. ½° W. This island is about six miles in length, and like most others in the gulf appears rocky, barren, and uninhabited by animals.

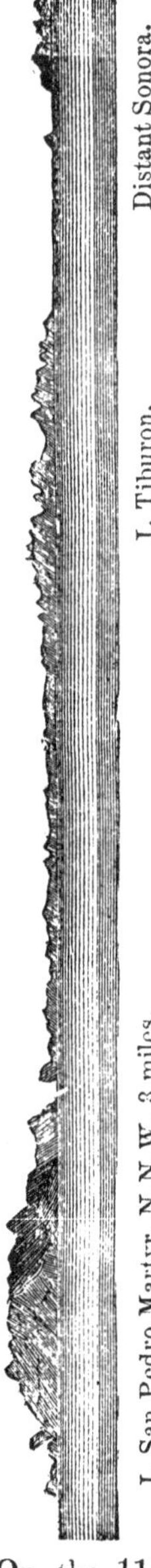

The violence of the wind abated on the 8th, and on the 9th we found ourselves off the large bay represented on the chart as lying between capes Trinidad and San Miguel, on the California shore. Here we observed with glasses a very extensive plain, covered with cactus. On the 10th we passed the island of "San Juan Pedro Martyr," the longitude of which we determined to be 111° 54′. The island is a barren rock, about five miles long, situated nearly in the middle of the gulf. Far beyond, to the northeast, we saw the rugged outline of the celebrated island of Tiburon. This island has long been known as the abode of the Ceres Indians, a small tribe of about five hundred, who are represented as extremely hostile, and invariably opposing any attempt at landing; they are said to be armed with poisoned arrows. A rich bed of pearl oysters is said to exist between this island and the coast of Sonora, and there are accounts of rich gold mines upon the island; but as no one is ever known to have landed there, it is difficult to understand how the fact was ascertained. We should have landed and attempted an examination of the island, but the wind being contrary compelled us to lay over to the California shore, and we passed it in the night.

On the 11th we passed the island of San Estavan, which presents an

Bearing N. N. E. 10 miles.

exceedingly wild and sterile appearance, resembling the generality of the gulf coast on the California side. The appearance of the water between this island and that of San Lorento led us to suppose that we were on soundings, but on trying the deep sea lead we found no bottom at one hundred fathoms.

On the 12th we passed near three small islands (to the south of the large island of Los Angelos,) which are not put down on the chart, and of which we had no previous information. They are composed of coarse clay slate, colored trap, and trap tuffa, and covered with cacti. Immense quantities of seals are found upon these, as well as most of the other islands in the gulf. I named this group "Allen's Islands," as a compliment to Major Robert Allen, chief quartermaster of the Division. The longitude of the group is about 112° 40′ W., latitude 28° 55′ N., and the soundings in the vicinity from four to seventeen fathoms.

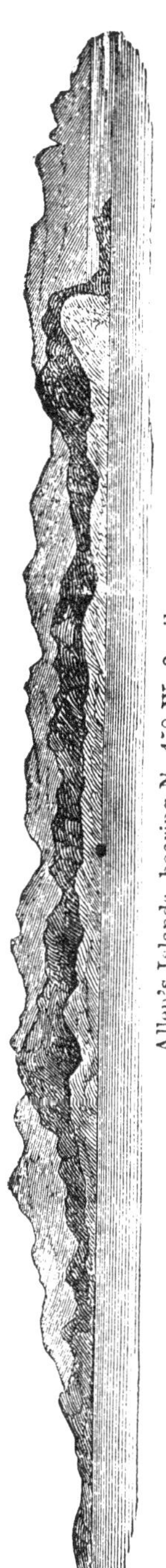

Allen's Islands, bearing N. 45° W. 3 miles.

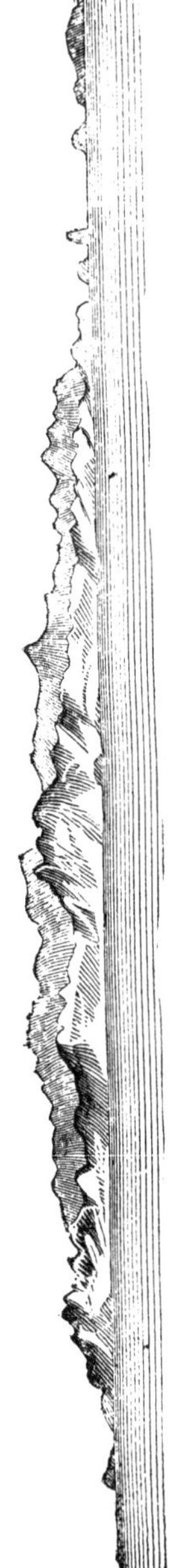

Angelos Island, N. W. 7 miles.

Angelos island, which we also passed on the 12th, is separated from the California shore by Whale's channel, which is about ten or fifteen miles wide. The island is about thirty miles in length, mountainous, rocky, and uninhabited. It lies between 29° 10′ and 29° 50′ N., and in longitude about 113° W.

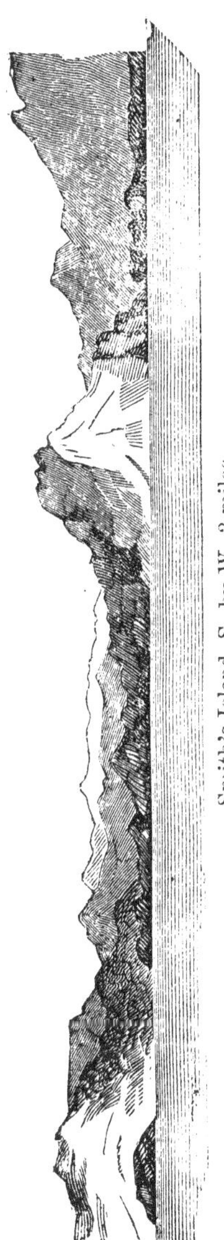
Smith's Island, S. by W. 3 miles.

We had received information in Guyamas of a small bay somewhere upon the California coast, in this vicinity, where fresh water could be obtained, and on the 13th, having rounded the northern extremity of Angeles island, we ran before the wind down Whale's channel, (where we observed a southerly current of about three miles per hour,) in search of it.

We discovered a large island lying close to the California shore, off the southern extremity of Angeles island, in latitude 29° 3′, longitude 112° 59′ W., which, not being put down upon any chart, I named Smith's island, in compliment to the general commanding, by whose order the expedition was undertaken.

This island is about a mile north of Angeles bay, which we found and entered on the 14th. This is a small indentation in the coast, about four miles in width, slightly open to the eastward, with a sandy bottom, in about ten fathoms water. The adjacent country is extremely wild and rugged. The hills are covered with huge boulders, which are coarsely aggregated masses of quartz, mica, and feldspar, and I noticed several isolated blocks of grey granite and sienite. There are three springs of slightly brackish water at the foot of the highest hill, and a bed of most excellent oysters is exposed at low tide. The water may be found by noticing the reeds which grow about it, and which are the only green things to be seen in the vicinity.

Angeles Bay.

The deep and rugged channels running down the sides of the mountains, and the immense pile of boulders forming their sides, would seem to indicate that at certain seasons of the year great quantities of water must fall here; but at the time of our visit everything appeared dry and parched. There are many varieties of cactus in the vicinity, but no nourishing herbs or grasses, and we saw no traces of game, but a few tracks of coyotes and wild goats near the spring. The bay is probably well known to the people inhabiting the interior, as we found many traces of old encampments, piles of oyster shells, heaps of ashes, and many mule-tracks leading to the southward. There are plenty of turtle in the harbor, but we did not succeed in taking any. The sketch will give an idea of the character and topography of the bay, though by no means the result of an accurate survey.

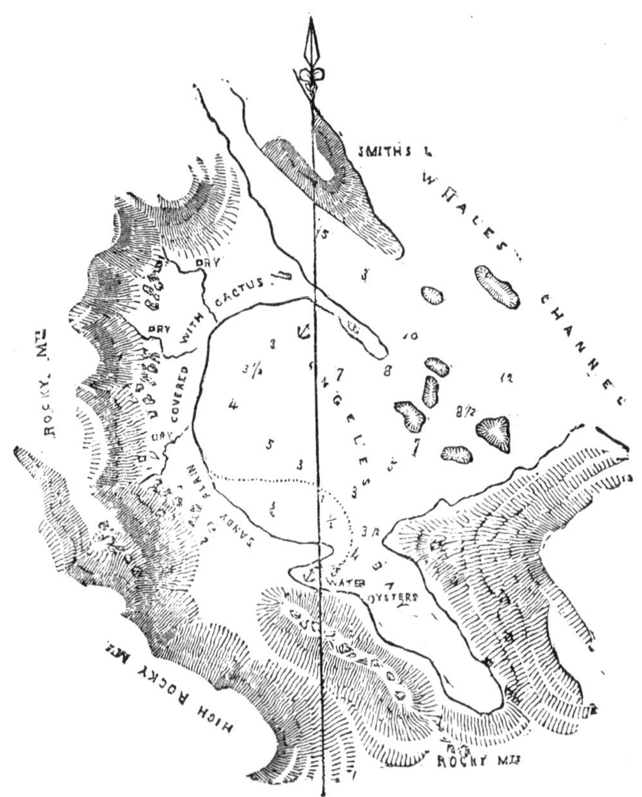

We remained at Angeles bay until the 18th, when, having filled our water-casks, we weighed anchor and stood out from our anchorage into Whale's channel. It being low tide we noticed several sunken rocks at the entrance, which we had not perceived on coming in. The rise and fall at Angeles bay is about fifteen feet. Its longitude is 113° 25′ 30″ W., latitude 29° 5′ N. We had rain on the 19th in squalls, and on the 20th a gale from the northwest, against which we steadily beat up for the head of the gulf. On the 21st we sounded when about five miles from the California shore, and found the depth from twenty to twenty-five fathoms. On the 22d we passed within four miles of a very remarkable rock, which at a distance has the appearance of a ship with all sail set. I named it the "Ship rock," but am inclined to think it identical with Hardy's "Clarence

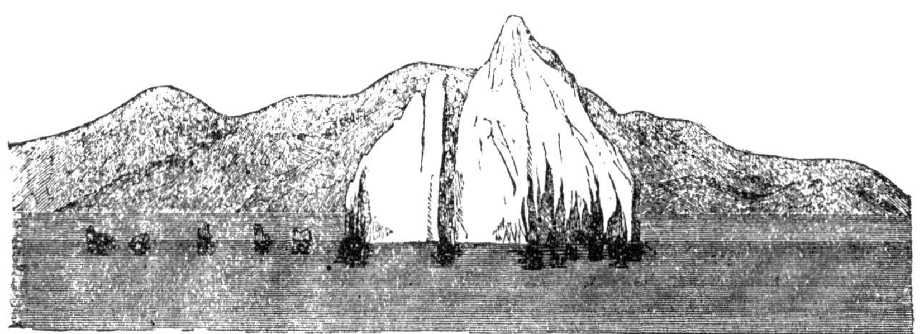

Ship Rock.

island," although twenty miles to the westward of the position which he lays down. We sailed over the latitude and longitude of his Clarence island, but saw nothing of it. The Ship rock is apparently a mile in circumference, about two hundred feet in height, and perfectly white with guano. We sounded continually on the 22d, and found a depth of twenty fathoms, with a sandy bottom. In the afternoon we observed a column of smoke in the distance, bearing N. 60° W., which we afterwards ascertained to be caused by the Indians burning the reeds on the river Colorado. On the 23d the land was plainly discernible on both coasts of the gulf; on the California side bold and mountainous, but on the Mexican low and sandy. Sounding continually, we found but about thirteen fathoms water, and we noticed constantly masses of reeds, trunks of trees, &c., floating past us, and at 6 p. m., the wind having died away, we dropped anchor in *six fathoms water.* We found by throwing the log, a current setting southward about a mile and a half per hour. The soundings from the 21st had shown us that the bottom was a fine blue pipe-clay, which (mentioned by Hardy) indicated our approach to the mouth of the river. On the 24th we weighed anchor and ran over to the California shore, the water shoaling gradually from six fathoms to ten feet, and at night we anchored between Montague island (which we readily recognised by Hardy's description, see chart) and the main land. The country here is low, flat, and covered with dwarf reed and coarse grass. Thousands of trunks of trees lie scattered over its surface as far as the eye can reach, showing that it must be entirely overflowed at the season of freshets.

We landed on the 25th, both on Montague island and the main, and found the soil clay detritus and the vegetation scanty. The two islands situated in the mouth of the river, called by Hardy Montague and Gore, are low, flat and sandy. They are separated by a channel about one mile in width, and extremely shallow. They are evidently formed by the accumulation of the sand and detritus from the river, and are gradually increasing in size. We commenced the survey of the river upon the 25th, which we continued from day to day, as we ascended. On the 27th, by taking advantage of the tide, we had succeeded in reaching Unwin's point, off which we anchored, and noticing the smoke of several fires in the distance, which we hoped might be at Major Heintzleman's encampment, fired several guns to warn him of our approach. The log gave us at this point a current at ebb tide of four miles and a half per hour, which we found to be the average velocity, except at spring tides, when it is much increased.

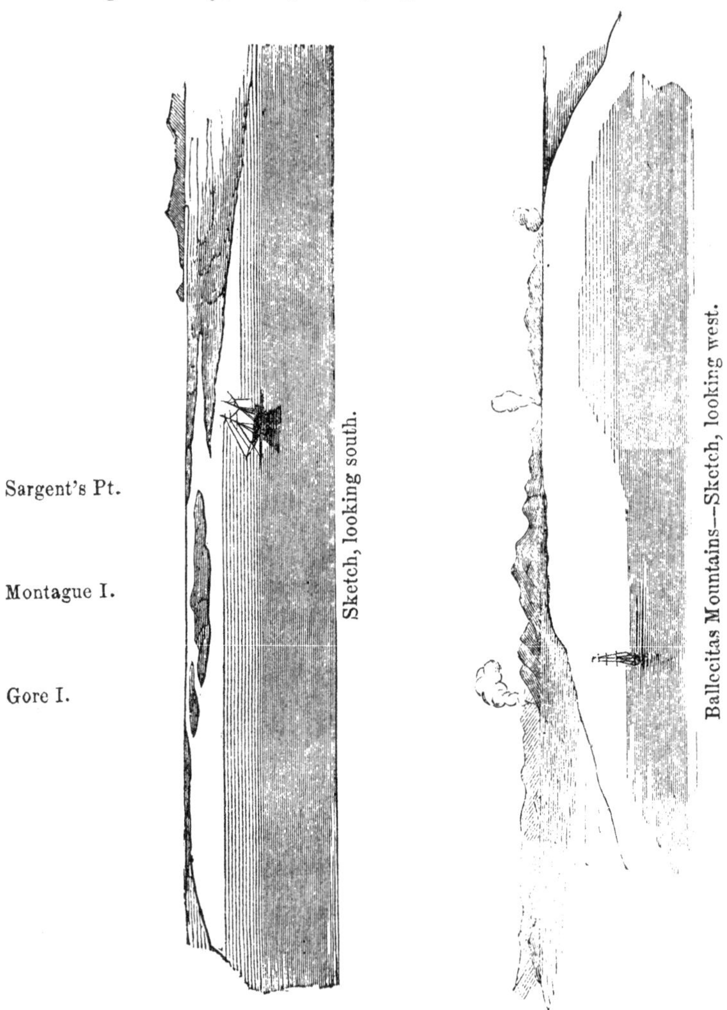

Sketch, looking south.

Ballecitas Mountains—Sketch, looking west.

On the 28th we left our anchorage at 7 a. m., and beat up with the flood tide until about 9, when we grounded off Charles's point. The soundings had been from one and a half to three and three-fourth fathoms. At this point we found the water devoid of any brackish taste, but extremely muddy, resembling in its character that of the Mississippi river. We continued daily up the river, sounding the channel at low water, and marking it by stakes, starting with the flood-tide and floating with it until the ebb set in, when we anchored and went on shore to continue the survey. The shores of the river (here from two to four miles wide) continue of the same character, the grass growing somewhat more thickly as we ascended, however; and upon "Green Hithe point," off which we anchored upon the 1st of January, we found a thick growth of artemisia. We had experienced no little difficulty in crossing from Charles's point to Green Hithe point; the channel is extremely narrow, and nearly at right angles to the direction of the river, and we had twice been swept from it by the tide and grounded upon the bank above. Upon the 1st, however, we were favored with a strong breeze, taking advantage of which we soon crossed and anchored in the channel upon the southern side. We landed upon Green Hithe point (formed by two small indentations in the crumbling clayey bank) and found the land on the south bank to be a perfectly level plain, (the soil clay detritus,) extending to the south and west for miles, and intersected by numerous sloughs, apparently filled by every spring tide. The river westward is bounded by the high hills of the lower California chain, at the distance apparently of fifteen or twenty miles. Trunks and limbs of large trees, some recently deposited, others in an advanced state of decay, are thickly scattered about, evidently left by the spring freshets. The river bank is abrupt and about twenty feet in height; the water is gradually undermining it, and large pieces are continually dropping off, falling into the water with a sound as loud and not unlike the distant report of a musket. We discovered two new islands in the reach above Green Hithe point; they are low and sandy, separated by a small channel and covered with coarse grass; they have undoubtedly been formed within a few years. The bed of the river is filled with quicksand, and its current at the time of spring tides is so strong that the channel is continually changing. I named these islands Gull and Pelican respectively, from the numerous flocks of these birds continually hovering or afloat in their vicinity. We discovered the remains of a rude Indian hut near the shore, and observed many tracks of horses in the vicinity. During the night of the 1st the vessel grounding at ebb tide, swung around upon her heel, and thumping violently was carried by the tide (dragging her anchor) some two or three miles, grounding finally upon the shoal of Gull island; at flood-tide sail was made on her as soon as she floated, and we succeeded in getting her back into the channel. As the vessel grounded at every ebb-tide, and on the return of the water was violently swung around, thumping on her bottom and swinging on her anchor, I began to see that it would be neither prudent, or in fact possible, to ascend the river much higher, and we accordingly commenced making preparation for a boating expedition; the tides were now on the increase, and on the 2d I observed, with the log, the velocity of the current at ebb-tide to be five and a half miles per hour. We observed on the 2d, for the first time, the singular phenomena of the tide coming in, in a bore or wave, while the ebb was still rushing past the vessel towards the gulf. On looking in the direction of Green Hithe point, a bank of water some four feet in height, extending clear across the river, was seen approaching us with equal velocity; this huge comber moved steadily onward, occasionally breaking as it rushed over the shoals of Gull and Pelican islands; passing the vessel, which it swung around on its course, it continued up the river. This phenomena was of daily occurrence until about the time of neap tides, and shows the truth of Hardy's assertion, that "*there is no such thing as slack-water in the river Colorado.*" I took the whale-boat on the afternoon of the 2d and proceeded up the river with the flood-tide, rounding Howard's point, (so called by Hardy;) we found ourselves in a broad but shallow bay about four miles in width. At the northeast and northwest extremities we found the two branches of the river, the former of which he mistook for the Gila; this is in fact the main channel of the river, the other being merely a slough which divides the river, about a mile from its entrance, into two branches, one of which terminates in a small lagoon about four miles from its mouth, the other communicating with the river above. As there is not water enough in either of these branches to float a whale-boat at low tide, it is evident that the river must have altered entirely since Lieutenant Hardy's visit, or that he never ascended it as he says he did with the Briga, a schooner of twenty-five tons. We sounded clear across the bay and found from two and a half feet to one and one-fourth fathoms. As the schooner was drawing eight feet, this settled the question as to her navigating the Colorado above this point.

On the 3d, having ascended the river with the schooner to within four miles of the point reached the day previous with the whale-boat, we fired three heavy guns at short intervals, still hoping that Major Heintzleman's command was in hearing distance of us. The Captain, Mr. Ferrier, and myself then went on shore and walked into the country, over the level plain in the direction of the nearest range of hills. The soil is a mixture of sand and alumina, evidently overflowed and sparsely covered with coarse grass, reeds and flag, with here and there tufts of musquite. We observed many more foot-prints, both of men and horses, and found the remains of another rancheria; after proceeding some three miles we heard a gun from the schooner, and turning saw the ensign hoisted at the main. We immediately returned and discovered the cause of the commotion in the approach of fifteen or twenty Indians, attracted by the report of our guns in the morning. On arriving at the beach they laid aside their bows and arrows, and made signs to us to approach, and addressed us in Spanish. We took six of them including their chief on board. They were very much like all other Indians, with coarse black hair, like a horse's mane, cropped straight across their eyes, and with no clothing except the inevitable dirty rag worn apron fashion about their loins; they were, however, a little nastier than any Indians I had ever before seen, being beplastered from head to foot with mud, with which some of them had filled their hair; one had on a linen coat, and another, to our great joy, an infantry great coat. After treating those on board with bread and molasses, I ascertained from the chief that they were acquainted with Major Heintzleman, and would carry a letter to him from me, which they said would take a day and a half; I accordingly prepared a letter, and having paid the chief in advance, sent him off with it.

We were frequently visited afterwards by these Indians and many others, to the number, perhaps, of two hundred; they call themselves the Co-co-pas, and live in a little village of from twenty to fifty inhabitants, dispersed about near the bank of the river; the men are very tall and strongly made as a general thing, and the women are modest, well behaved, and rather good-looking; their huts are precisely like those of the California Indians, made of sticks in a spherical or oven shape, and covered with dirt. They live on fish, small game, and bread made of grass seeds, and raise pumpkins, watermelons, &c., on little patches of ground which they cultivate; we found them very friendly, quiet and inoffensive; they brought fish to sell to us nearly every day, and though continually on board the vessel, we never missed even the most trifling article. Their arrows are made of reed with a pointed end of hard-wood, and their bows of willow, so that if they were disposed to be as hostile as had been falsely represented to us, they would be incapable of doing much damage. They were annoying visiters, however, pervading every part of the vessel, gazing at every thing that was going on, begging for shirts and other articles of clothing, and afflicted with a never-to-be-satisfied hunger. It was remarkable that although they had never seen a vessel before larger than their own reed rafts or balsas, they never expressed the least surprise at the "Invincible," or anything connected with her, appearing to take our arrival as a matter of every day occurrence. The villages higher up the river own many horses, and the chiefs always rode when coming to visit us. I suppose the whole tribe may number one thousand, including men, women and children. The men frequently wear beads, rings, &c., in their noses, and paint their faces black and red with charcoal and ochre. I sent off two other Indians with letters to Major Heintzleman, reporting our arrival, and having waited without a reply for more than a week, during which time the carpenter had been experimenting upon the long boat, putting a shaft with cranks and paddles in it, (which when tried proved of course a signal failure,) I concluded to continue the ascent of the river with a boat; the Indians who had visited us had frequently said that higher up the river lived a bad race of Indians who were thieves and murderers, (we found afterwards they meant the Yumas on the Gila;) I therefore had the long-boat armed with one of the swivels, and taking provisions for ten days, started on the 11th with the Captain, Mr. Ferrier, and six men for the junction of the Gila. We reached Arnold's point (between the main channel of the Colorado and the western slough) about 9 a. m.; here a new island one-fourth of a mile square has been formed, but at present is covered at spring-tide. We rowed up the main channel (Hardy's Gila) about eleven miles, encamping at sundown on the bank. We found the river very winding, the depth of water in the channel varied from four to fifteen feet, and the action of the tide ceases to be discernible five miles above Arnold's point, or about forty miles from the mouth of the river. The banks are lined with rushes, cane, small willows and acacia, and occasionally we observed small cotton woods or poplars; the river was full of geese and ducks, and on the banks we observed may deer tracks of unusual size. The width of the river varies from two hundred yards to half a mile; we saw no Indians, though they had set fire to the rushes in our vicinity, and we noticed two of their rush balsas lying on the beach. The banks of the river were generally from ten to twenty feet high, of stratified clay. On the morning of the 12th we continued our voyage, passing about four miles above our encampment to an Indian village, the inhabitants of which to the number of one hundred, among whom I recognized some old acquaintances, sat upon the bank watching our progress in silence. We passed another rancheria in the afternoon, near which lay an old scow built of wagon bodies, which had floated down from the ferry. We encamped at sundown, having made about twelve miles. The banks we observed in many places were of fine white sand. We saw a great deal of game during the day, and we found a well-trodden horse-trail running north and south about a mile in rear of our encampment. The velocity of the current on the river had been about one and a half and two miles per hour. Three Indians came up to our encampment from their village during the evening and presented us with a few pumpkins; they returned during the night. The banks were more thickly wooded than below, but we saw no large trees; we observed great numbers of bear and other tracks upon the banks. We started at sunrise on the 13th, and after pulling steadily forward about five miles as we rounded a bend, I observed at a distance a white object which I took for a house, but the splashing of the water by its sides glittering in the sun soon showed it was a boat. We fired our swivel immediately, and had the pleasure of having our salute replied to by two muskets. It was Major Heintzleman, who, accompanied by Dr. Ogden and Mr. Henchelwood, (proprietors of the ferry) had left the post on the 11th, on receipt of my letter, and according to their estimate had descended the river over eighty miles. As the major had made field notes of the river from the time of leaving the Gila, I considered it unnecessary for me to proceed further, and we accordingly pulled down the river in company. We reached the schooner that afternoon, and on the 14th succeeded in getting her about three miles further up the river to a point two miles southeast of Arnold's point on the eastern side. Major Heintzleman left on the 15th, promising to send a party down immediately on his return to take charge of the provisions. On the night of the 16th (spring-tide) the vessel surged so violently on her chain as to part it, and we lost our smaller anchor; and on the 18th the flood-tide wave coming in with great velocity, struck the vessel or her counter as she lay aground with great violence, carrying her around with it with such force as to snap the ring of the large anchor, and we found ourselves drifting rapidly up the stream towards Arnold's point. Our situation was one of great peril, for this being the highest spring-tide, if we had grounded above, the vessel must have been lost. But the captain with admirable coolness and self-possession seized the helm, and directing her towards the bank, ordered the men to leap from the jib-boom on shore with the kedge anchor, which fortunately brought her up.

We then procured the largest logs that we could find, to which, when sunk in the ground, springs were attached from the vessel, and rigged out spars to keep her at a safe distance from the bank, thus rendering her as secure as possible; our situation was, however, a very unenviable one, as, apart from any danger that might arise in our present position, we had to dread the voyage down the river, which, without anchors, and liable continually to get aground, threatened to be extremely dangerous; the accident was, moreover, to be regretted, as preventing us from touching upon the California shore in search of harbors, and at the island of Tiburon, as I had intended; on our return, every exertion was made to recover the anchors, but without success.

We remained in this position until the 27th of January, without incident, occupying the time in exploring the western slough, putting up channel marks in the river, &c. We were visited almost daily by crowds of Indians, some of them from a considerable distance up the river; they generally brought fish with them to sell, and conducted themselves in the most friendly manner. The river appears to be most plentifully stocked with fish, principally salmon, trout and suckers; a species of perch is sometimes found; all the fish have an unpleasant, muddy taste, and their flesh is soft and watery.

Lieutenant Murray arrived from Major Heintzelman's camp upon the Gila, on the 27th; he was accompanied by Dr. Ogden and a sergeant with eight men; they brought with them a wagon, and stated that the bank, for three miles above, was muddy; but as this was caused by the late overflow of the spring tides, I concluded that the ground would settle before the full moon, (15th February) and that the supplies might be safely landed; accordingly, upon the 28th, we landed all the provisions upon the bank, piling them up carefully on logs, and digging a ditch around them, and pitched tents for the men who were to remain with them. On the 29th we cast off from the bank, and floated down the river, taking advantage of the ebb tide. We got aground twice in going down, but fortunately got off at high water without accident, and on the 1st February had crossed the bar at the mouth of the river, and were fairly in the Gulf of California on our return. The bar at the mouth of the Colorado is about ten, possibly fifteen miles in width; the soundings upon it are from ten feet to four fathoms; it is a very loose, muddy bottom, and with a stiff breeze a vessel could force her way over it, even if drawing a foot or more than the lead would indicate. The distance from the junction of the Gila and Colorado to the mouth of the latter, by water, is about one hundred and four miles, owing to the many bends of the river, though the difference of latitude is but little more than half that distance. The navigation of the Gulf of California presents none of those difficulties which we had been led to anticipate. The wind we found invariably from the north-west, which, at this season of the year, is its prevailing direction; it is only during the months of June, July and August that the gales from the south-east are prevalent; except in Whales' channel we noticed none of the strong currents so frequently mentioned as existing in the gulf. It would be difficult to mistake the entrance to the river, it being in fact the head of the gulf, which gradually narrows from forty to three miles when it is joined by the river, whose turbid stream discolors the gulf for many miles to the southward, in soundings of twelve and fourteen fathoms. On the Sonora coast, however, exists an indentation some fifteen or twenty miles in depth, called by Hardy, Adair bay; the shoals of this can be seen from the mast head, a view from which would prevent one falling into the error which he did of supposing it a mouth of the Colorado. The angle at the junction of the slough and the main river is called Arnold's point, and from the mouth of the river (after crossing the bar) to this point, the channel varies in depth from fifteen to thirty feet, at ordinary high tide, and may, as we have practically demonstrated, be ascended by a vessel having a draft of eight or nine feet, by taking advantage of the flood, which has a velocity of from three to five miles per hour. It is impossible to sail up, however, for although the river varies in width from three miles to six hundred yards, the channel is narrow and the navigation elsewhere obstructed by the numerous sand bars. The proper method, after passing Gore and Montague islands through the western channel of the river, is that which we adopted, to drift with the flood tide, keeping close to the highest bank, sounding continually, and anchoring before the time of high water: in this way we progressed slowly but steadily; making, perhaps, four or five miles per day, until we arrived at the point where we finally landed the stores, which I have named, as will be seen by reference to the map, "Invincible point." Above Arnold's point the river is very circuitous, the swell of the tide rapidly decreases, the channel becomes narrow, and the water has less depth. At this season, therefore, Arnold's point may be considered the head of navigation for vessels of nine feet draught; above this point we found always from three to fifteen feet of water in the channel, whose width varied from fifty to three hundred yards; and as the river at that time was at its lowest stage, I have no hesitation in saying that it may be navigated at any season of the year by a steamboat of eighteen or twenty feet beam, drawing two and a half to three feet water. A small stern-wheel boat, with a powerful engine and thick bottom, I would respectfully suggest to be a proper description of vessel for this navigation, where a strong current has to be contended with, and the channel, somewhat obstructed by small snags and sawyers, is quite narrow in several places. At the present season (January, February and March) supplies from vessels arriving from the gulf may be landed near Arnold's point, upon the eastern bank, and a road being made from the post (a work of little difficulty over a level sandy plain) they might be transported by wagons across in three days. It would be preferable, however, to establish a depot by anchoring a hulk near Charles point, laden with stores, from which a small steamboat could carry more to the post in twenty-four hours than a hundred wagons could transport in a week. Either of these methods would be far preferable to the present slow, laborious and uncertain mode of supplying by wagons and pack mules across the desert from San Diego.

The time of high water at "full and change" at Arnold's point is 3h. 20m. p. m., and the rise of ordinary spring tides about twelve feet; but during the season of freshets the river throughout its extent (judging from the statement of the Indians and the indications upon the banks) is at least fifteen feet higher than at the time of our visit, and the velocity of the current which, above the effect of the tide, was from one to three miles an hour, is nearly doubled; the action of the tide ceases about forty miles from the mouth; the banks of the river are low, flat, and either sandy or of crumbling clay which appears to have been deposited in successive strata. Near the mouth there is no vegetation, but higher up the shores are thickly lined with cane, rushes, small willows, acacia and cotton wood, and the country in the interior covered with a coarse sharp grass. The latitude of Invincible point is 30° 50′ 26″, its longitude 114° 46′ 43″ W. The subjoined sketch, on a larger scale than the map, will give a good idea of the topography of the vicinity of Arnold's Point.

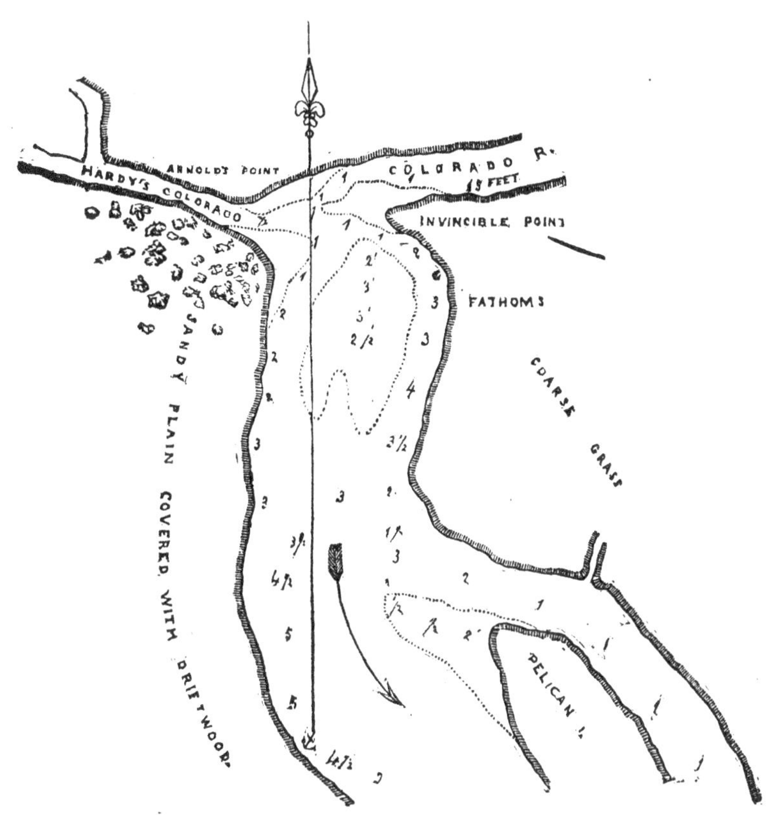

The large shoal south of the point is an island, except at spring tides, and is gradually increasing in size and height.

We reached Guyamas on our return on the 5th of February, and running into the harbor before a light breeze, let go the kedge anchor with sixty fathoms of chain, which we found would hold the vessel in still water. The British frigate Meander was lying in the harbor, and noticing as we passed her that we were without anchors, her commander, Captain Keppel, R. N., very handsomely sent a boat alongside us with every offer of assistance. We had, however, already accepted the kind offer of Mr. Southworth, owner of the schooner Mary Phebe, of San Francisco, who lent us one of his anchors until we could purchase one; this we succeeded in doing at a reasonable rate, and finding every thing in the way of fresh provisions exceedingly high in consequence of the presence of the frigate, we sailed upon the 9th for San Jose upon the California coast. I compared my chronometers with those of the Meander while at Guyamas; and experienced from her officers every kindness and attention in their power to offer. We arrived at San Jose on the 11th. We found the roadstead open to the south from E. N. E. to S. W.; it is unsafe during certain seasons of the year; the bottom is very uneaven, but the holding-ground good, being white sand. We found on our arrival the ship Glasgow, from Liverpool, and five coasters lying at anchor. The mission is situated about a mile from the beach, upon a small stream which, watering the valley of San Jose, divides into four small branches and empties into the sea at this point. The valley extends north some fifteen miles and is environed by the high and rugged mountains of the lower California range. The population of San Jose is between five and six hundred. A large number of Americans, I was informed, had settled in the valley above. The soil appears sandy, which is owing to a severe flood that occurred about eighteen months ago, washing away all the upper and richer surface into the sea, and forming the four branches to the stream which we noticed. Grapes were previously abundant; now there are but two or three vineyards. The sugar-cane and banana are cultivated here as well as sweet potatoes, pumpkins, Indian corn, &c., but the crop this season will be small, the inhabitants having abandoned their gardens during the working season and gone to the interior to avoid the cholera. Cattle, mules, swine and goats are bred in the vicinity, but of the first many perish on account of the drought which had been unusual, and was severely felt in the interior. I noticed here great numbers of a bird called the secretary. We sailed from San Jose on the 12th, having procured some vegetables of which we stood much in need, and after a long voyage of twenty-one days without any incident worthy of recording, arrived at San Francisco on the 6th day of March.

I have the honor to transmit herewith an extract of my journal kept during the voyage, showing the range of the thermometer, state of the weather, direction of the wind, and geographical position of the vessel for each day during the expedition. Also, a chart of the gulf and river Colorado, from its mouth to the junction of the Gila. The river is platted from my own notes as far north as Heintzleman's Point, above this from notes made by Major Heintzleman. The latitudes given are calculated from meridian altitudes of the sun, taken by Captain Wilcox, Mr. Ferrier and myself; the longitudes from a daily mean of nine observations taken for the error of our chronometers near 9 a. m., and when possible, as in port, from equal altitudes of the sun in the forenoon and afternoon.

I am, sir, with high respect, your obedient servent,

GEO. H. DERBY,

Brevet 1st Lieut. Topographical Engineers.

Lieut. Col. J. Hooker,

Adjutant General Pacific division.

Abstract of a journal kept on board the United States transport Invincible, during her expedition to the Gulf of California and the river Colorado, from November, 1850, to March, 1851, by Geo. H. Derby, Lieut. Topographical Engineers.

Date.	Latitude.	Longitude.	Thermometer	Direction of wind.	Remarks, embracing weather, position of the vessel with reference to islands, land in the vicinity, &c.
1850.	° ′ ″	° ′ ″	S.R. M. S.S.		
Nov. 1	37 48 00	122 21 00	....	West breeze................	Sailed from San Francisco at 1 p. m.; south point of entrance bore north, distant 5 miles.
2	35 23 21	122 30 00	....	Northwest to west...........	Thick weather, Santa Barbara at noon bearing south-southeast 8 miles; heavy rain during the night.
3	32 34 00	119 56 45	....	West and changeable........	9 a. m.; Island St. Clements bearing north 5 miles.
4	32 50 00	118 09 15	....	East to northeast by north....	Clear weather at 10 p. m.; Coronadas Island bearing east-southeast; anchored off Beach San Diego at 11 a. m.
5	32 50 00	118 09 15	....	............do.............	Discharging; took Acting Assistant Quartermaster Johns on board and landed at New San Diego.
6	32 50 00	118 09 15	....	Moderate from west to northwest.....................	Pleasant weather; discharging cargo and repairing sails at New San Diego.
7	32 50 00	118 09 15	....	............do.............	Taking on board part of the same cargo to land at the beach or Plaza.
8	32 50 00	118 09 15	....	............do.............	At new town of San Diego.
9	32 50 00	118 09 15	....	............do.............	At 4 p. m. sailed for the Plaza, arriving at 8 p. m.
10	32 50 00	118 09 15	....	............do.............	At the Plaza discharging cargo.
11	32 50 00	118 09 15	....	............do.............	Repairing schooner's copper on the ballast spit below the Plaza.
12	32 50 00	118 09 15	....	............do.............	do. do. do.
13	32 50 00	118 09 15	....	............do.............	do. do. do.
14	32 50 00	118 09 15	....	............do.............	Returned to the Plaza and received on board supplies for the Gila.
15	32 50 00	118 09 15	....	North-northwest	Sailed at 3h. 30m. p. m. with good breeze; weather cloudy.
16	32 08 00	No obs.	....	Very unsteady...............	Weather cloudy.
17	30 41 00	117 52 15	68 73 72	West to northwest...........	At 5 p. m. island of Gaudaloupe bearing southwest half south, distant about 30 miles; weather fine.
18	28 46 00	117 25 45	69 72 70	Northwest to west...........	Clear and fine.
19	27 24 00	116 54 30	68 76 70	North-northwest	Saw the Alijo rocks at 11.30 a. m., bearing south-southeast, about 10 miles distant.
20	25 07 00	115 42 00	70 78 71	North-northwest to north.....	Clear; Alijo rocks still in sight.
21	22 50 00	112 26 00	72 85 74	North......................	Cape St. Lucas at 6 a. m., bearing east-northeast.
22	22 38 30	110 48 00	80 86 80	North—very light...........	At noon Cape St. Lucas bore northeast by east, Cape Falso north by west.

ABSTRACT—Continued.

Date.	Latitude.	Longitude.	Thermometer	Direction of wind.	Remarks, embracing weather, position of the vessel with reference to islands, land in the vicinity, &c.
1850.	° ′ ″	° ′ ″	S.R. M. S.S.		
Nov. 23	22 46 43	110 04 30	82 86 82	North-northwest and north...	4 p. m. passed Cape St. Lucas; put about and anchored in the bay within 200 feet of beach; sailed at 6.
24	22 46 32	108 43 00	80 84 80	North....................	At noon Cape Palmo, west by south, 30 miles.
25	23 27 41	108 49 30	78 82 79	Southeast by east—strong....	At noon Island St. Ignacio, bearing east-northeast, distant 10 miles.
26	25 24 43	100 26 00	70 72 78	Southwest by west..........	Strong gale; at 11 a. m. close in to the shore of Lower California.
27	26 29 00	110 53 00	67 70 67	North-northwest	Strong gale 11 a. m.; Lobas island bore east-northeast, distant 6 miles.
28	27 18 45	110 35 30	68 72 70	Northwest................	Strong gale; at noon Cape Haro bore northeast, distant 10 miles.
29	27 43 11	110 59 00	70 76 70	North-northwest	Strong breeze; at 4h. 30m. p. m. anchored in Guyamas bay in 3½ fathoms.
30	27 53 05	110 49 13	70 77 72	Little or no wind...........	Watering and repairing.
Dec. 1	27 53 05	110 49 13	72 77 72	.......do....	Do do.
2	27 53 05	110 49 13	68 75 73	.......do....	Do do.
3	27 53 05	110 49 13	66 70 67	.......do....	Do do.
4	27 53 05	110 49 13	60 68 66	.......do....	Do do.
5	27 53 05	110 49 13	60 65 62	North-northwest	Heavy gale; sailed at 3 p m., passed Cape Haro at 4, at 6 island of Rolasco bore northeast by east half east; shivered the topsail.
6	27 56 00	111 6 45	61 66 60	North-northwest	Strong gale; at noon Rolasco bore east-northeast, made two points leeway, sea very high.
7	27 58 18	111 39 00	62 68 62	Northwest by north..........	Clear weather; very high sea, split the flying jib.
8	27 58 23	111 23 00	69 72 62	Northwest by north..........	Weather began to moderate.
9	28 15 00	111 43 30	68 71 68	North.....................	9 a. m. island of San Pedro bore northwest by west 3 miles.
10	28 22 00	111 59 00	67 73 70	Light.....................	At 10h. 30m. a. m., island of San Estefan bore north-northeast 8 or 10 miles.
11	28 27 00	112 21 30	68 72 70	North.....................	At 1 a. m. south of island of Lorento; 6 p. m. sounded at southwest of Lorento, no bottom, 100 fathoms.
12	25 51 46	112 45 00	68 68 68	East......................	2 p. m. ran through small group named Allen's islands; soundings 17 fathoms, Angeles island northwest 8 miles.
13	29 07 00	112 59 00	68 70 66	Light and variable..........	Beating about Whale's channel for the bay of Angeles; current 3 miles per hour, southerly.

Date	Latitude	Longitude	Thermometer	Wind	Remarks
14	29 04 50	113 28 45	66 71 65	Northwest, with strong gusts through the mountains.	Anchored at north end of Angeles bay in $2\frac{1}{2}$ fathoms.
15	29 04 50	113 28 45	66 70 64	......do........do.........	Anchored at south end of Angeles bay in 4 fathoms; found water.
16	29 04 50	113 28 45	67 69 62	......do........do.........	Watering; the spring is about 800 yards from the beach at the base of the highest mountain; slight shower.
17	No observations........		65 67 64	North-northwest, strong......	Weighed anchor at 6h. 30m. and ran northeast by north through the islands at the mouth of the bay.
18	No observations........		64 65 63	Changeable and squally......	Beating about between Smith's island and the main land.
19	No observations........		63 67 63	Northwest, strong...........	Thick weather.
20	30 12 00	113 30 00	62 65 62	Northwest	A gale; clear; observed column of smoke to the northward.
21	30 43 49	113 51 00	62 66 64	North-northwest	Water muddy; about 3 p. m. sounded 20 fathoms, bottom white clay; 6h. 45m. off "Ship Rock" northwest 4 miles.
22	30 10 12	113 53 30	60 66 63	North-northwest	Beating up; sounding 19 fathoms; shoaling toward either side of the gulf; observed smoke north.
23	31 34 00	114 19 00	59 64 63	Southeast, light.............	At 8 p. m. dropped anchor in $3\frac{1}{2}$ fathoms; bottom soft mud.
24	31 41 00	114 9 30	54 53 60	Northwest breeze...........	Under weigh at 1 p. m., beat up to the mouth of the river; entered at 5 p. m.; soundings from 2 to 5 fathoms; missed the channel and got aground in soft mud; at 6 p. m. got off and anchored in 5 fathoms water.
25	31 41 00	114 9 30	50 60 60	Northwest by north..........	Head wind; went ashore and commenced survey; landed on Montague Island and the main.
26	31 51 23	114 10 45	44 64 59	Calm.........................	At 7 a. m. weighed anchor and drifted up with a three-knot tide; rounded the shoal at Unwin's Point, and anchored.
27	31 51 23	114 10 45	48 60 60	Northwest..................	Fired large gun; aground at low water; left anchorage at 7 and beat up with flood; grounded at Charles's Point at 9 a. m.
28	31 51 15	114 28 15	50 62 60	Northwest, strong...........	Started about 9 a. m. going up with the flood, but grounded at 11; high and dry at low water; found a deserted hut near the water's edge; continued survey.
29	31 51 15	114 28 15	52 64 59	Northwest, strong...........	Discovered two large islands in the river opposite Greenhithe Point; warped off and grounded again on Pelican Island shoal.
30	31 51 15	114 28 15	50 64 60	Northwest, mild.............	High water 1h. 40m. p. m., but insufficient to float the vessel; staked out the channel and continued the survey.
31	31 51 15	114 28 15	52 62 52	North northeast breeze.......	Got off at high water, but grounded afterwards on Greenhithe Point; continued survey.
1851. Jan. 1	31 51 15	114 28 15	53 61 56	Northwest, strong...........	Weighed anchor at 12h. 30m. p. m. having discharged part of our ballast, reducing the draught to 7 feet; anchored at 2h. 20m. p. m. in $2\frac{3}{4}$ fathoms; strong breeze right ahead; anchor dragged about three quarters of a mile during the night, and the vessel struck heavily, lying broadside on to the current.

ABSTRACT—Continued.

Date.	Latitude.	Longitude.	Thermometer	Direction of wind.	Remarks, embracing weather, position of the vessel and reference to islands, land in the vicinity, &c.
1851.	° ′ ″	° ′ ″	S.R. M. S.S.		
Jan. 2	31 51 15	114 28 15	52 60 57	North northwest, strong	At 12h. 35m. flood came up in a "bore" with a wall about four feet high, velocity five miles per hour by log; raised anchor and succeeded in beating up about three quarters of a mile, when we anchored (being head wind) and taking the boat went up the river to Hardy's Colorado, which we found a slough without current, the main channel being to the eastward.
3	37 49 26	114 36 30	53 62 58	Northwest, moderate	Continued survey; Indians arrived on the bank, about thirty in number; despatched the chief with a letter for Major Heintzleman.
4	37 49 26	114 36 30	60 64 54	South	Time of high water 4h. 30m. p. m; daybreak dull and misty; cleared up between 9 and 10 a. m.
5	37 49 26	114 36 30	55 63 55	Northwest, strong	Lying with both anchors out.
6	37 49 26	114 36 30	52 61 55	Calm	Rise of tide nine feet.
7	37 49 26	114 36 30	54 60 57	East	Foggy weather; continued survey.
8	37 49 26	114 36 30	53 62 56	Northwest, a gale	Clear; continued the survey.
9	37 49 26	114 36 30	54 63 58	Northwest	Commenced preparations for ascending the river in a boat, hearing nothing from Major Heintzleman, and fearing to take the vessel higher up.
10	37 49 26	114 36 30	54 74 60	Northwest	Cool damp morning; started with long-boat for the Gila; made at noon about 11 miles, having conducted a rough survey of the river, &c.
11	37 49 26	114 36 30	54 72 59	Northwest	Strong breeze and fine clear weather; camped about 15 miles above vessel at 6 p. m.; started at 7 a. m.; cool morning.
12	31 49 26	114 36 30	52 76 60	Northwest	Pleasant; encamped at 6 p. m. having made about 10 miles; current from one to three miles per hour; started at 7 a. m. and met Major Heintzleman at 9 and returned down the river.
13	31 51 00	114 36 30	54 68 54	Northwest	At 3 p. m. arrived at vessel accompanied by Major H.; weighed anchor and proceeded up the river one and a half mile.
14	31 51 00	114 36 15	53 62 60	Southeast breeze	Major Heintzleman started on his return.
15	31 51 00	114 36 15	54 65 60	South-southeast	Weighed anchor and made about two miles up the river, gaining a position near Arnold's Point.

	16	31 52 26	114 33 00	52 30 58	**Northwest, strong**	**Both anchors out; at 11h. 30m. p. m. lost small anchor, the chain breaking; flood very strong.**
	17	31 52 26	114 33 00	40 64 58	Northwest	Sweeping for anchor; at 8 a. m. lost large anchor, the vessel surging heavily on her chain against the flood-tide. Secured to the east bank by springs from bow and stern attached to heavy piles.
	18	31 52 23	114 33 00	55 67 60	Northwest	Fine weather.
	19	31 52 23	114 39 00	56 66 62	Norrhwest	Clear; dragging for anchors.
	20	31 52 23	114 39 00	58 67 60	Northwest breeze	Fine clear weather.
	21	31 50 02	114 39 00	60 70 64	Northwest	Clear.
	22	31 50 02	114 39 00	60 65 63	Northwest—strong	Clear.
	23	31 50 02	114 39 00	60 65 62	do	Cloudy.
	24	31 50 02	114 39 00	62 70 63	Calm	Cloudy.
	25	31 50 02	114 39 00	62 70 70	Calm	Dull and cloudy with slight showers.
	26	31 50 02	114 39 00	60 71 68	Southeast—light	Clear weather.
	27	31 50 02	114 39 00	62 72 69	North-northwest—light	Lieutenant Murray arrived after five days' march; distance by his estimate to Camp Yuma about 80 miles.
	28	31 50 02	114 39 00	63 75 72	do	Supplies all landed at 2 p. m.; Lieutenant Murray returned, leaving a guard of four men; clear.
	29	On river Col	orado	63 75 72	Northwest	Clear; sailed at 2 p. m. and grounded off Greenhither point.
	30	On river Col	orado	65 75 72	Calm	Clear and fine; vessel aground.
	31	On river Col	orado	60 72 70	North-northeast	Clear; got off, but was driven back by the flood-tide, grounding on the shoals of Pelican Island.
Feb.	1	30 06 19	113 08 25	63 73 72	Northwest breeze	1 p. m., floated and under way; 1·30, off Charles's Point; 4.30, off Sargent's Point; 12, in the Gulf.
	2	28 21 21	111 37 00	60 70 70	North-northwest—light	Angelos Island, bearing south half west; clear weather.
	3	27 55 21	111 7 45	61 70 69	Variable	San Pedro Martyr at noon, bearing west, distant 3 miles.
	4	Off ent. to	Guyamas	63 75 73	Calm	Beating about off Guyamas bay.
	5	27 53 50	110 49 13	63 73 72	Northwest to northeast	At anchor in Guyamas bay; clear and fine; (borrowed anchor from schooner Mary Phebe, San Francisco.)
	6	27 53 50	110 49 13	65 73 70	Southeast—variable	Guyamas; procured an anchor.
	7	No observa	tions	64 72 69	Southeast breeze	Weighed anchor and beat down the harbor to the ballast beach; heavy rain.
	8	26 15 08	111 01 00	60 70 67	Northwest breeze	At noon anchored at ballast beach; at 4 p. m. weighed and stood out; clear.
	9	24 20 01		64 68 64	North-northwest breeze	Fine weather; coasting along Lower California shore.
	10	No observa	tions	62 70 64	North-northeast to north-northwest.	8 a. m., Cape Palmo, distant 8 miles, bearing north.
	11	23 00 37	109 05 00	64 74 70	Variable	Anchored off San Jose in 16 fathoms; clear and fine.
	12	22 43 40	109 08 00	66 70 69	North-west—light	Sailed from San Jose; clear and fine.

ABSTRACT—Continued.

Date.	Latitude.	Longitude.	Thermometer	Direction of wind.	Remarks, embracing weather, position of the vessel with reference to islands, land in the vicinity, &c.
1851.	° ′ ″	° ′ ′	S.R. M. S.S.		
Feb. 13	22 43 40	111 56 00	65 71 70	Northwest breeze	Pleasant; Cape St. Lucas at noon, bearing west 15° north, 8 miles distant.
14	22 46 28	113 05 45	65 70 70	North	Cloudy with light wind.
15	22 40 00	114 35 00	66 70 68	North	Cloudy.
16	22 45 00	114 30 00	66 68 66	Calm	Clouded sky and a heavy westerly swell.
17	22 55 00	115 56 10	66 69 68	North	Moderate breeze, sky still cloudy.
18	23 21 45	117 24 00	67 69 68	North-northwest	Cloudy.
19	24 18 12	119 24 15	66 68 68	Northeast to north by east	Cloudy.
20	24 50 20	121 25 15	66 68 68	North	Cloudy.
21	25 25 13	123 14 30	66 67 66	North to north-northwest	Heavy sea; fair.
22	26 39 15	125 01 15	63 67 66	North	Strong breeze and heavy sea.
23	27 32 35	126 42 45	63 64 64	North by east	Light showers.
24	28 02 22	128 35 30	63 64 63	North by east	Cloudy and light showers.
25	29 40 05	129 34 00	64 66 67	North	Breeze; cloudy weather.
26	31 18 24	129 11 45	61 63 62	Northeast by east	Wind unsteady.
27	32 57 23	129 00 15	60 62 61	Northeast by east	Cloudy.
28	34 20 41	129 01 15	61 62 62	Northeast by east	Cloudy and hazy.
Mar. 1	34 58 29	127 59 00	60 61 60	West-northwest	Clear.
2	35 48 00	124 04 15	58 58 58	West-northwest	Foggy.
3	36 48 00	122 16 30	55 56 55	West-northwest	A heavy gale; thick weather; at 10 a. m. made land 45 miles south of Point Lobos.
4	37 10 00	122 07 00	54 55 55	Northwest	Strong breeze and heavy sea; clear.
5	....	....	55 55 54	Northwest	Light breeze and heavy fog.
6	San Francis	co..........	54 56 56	Calm to west	3 p. m. took pilot on board, and at 4.30 p. m. anchored in San Francisco bay.

www.ingramcontent.com/pod-product-compliance
Lightning Source LLC
LaVergne TN
LVHW011139110826
845150LV00008B/2408

9781418193003